P9-DMU-870

Cranberries

FRUIT OF THE BOGS

The publisher would like to pay tribute to the memory of Ken Nelson (1932–1994), who designed this book.

Cranberries

FRUIT OF THE BOGS

by **Diane L. Burns**

photographs by
Cheryl Walsh Bellville

Carolrhoda Books, Inc./Minneapolis

Special thanks to the following persons and organizations: Ralph May, Bruce May; Ocean Spray Cranberries, Inc., Kris Adams Wendt, Wendy Robinson, Walt and Doris Goldsworthy, and The Wisconsin State Cranberry Growers' Association; author/reviewer Cris Peterson, Three Lakes High School teacher Verdun Cecil, and DNR Water-quality biologist Bill Jaeger; The Cranberry Expo. Ltd.; Nicolet College Native American Center; Great Lakes Indian Fish and Wildlife Commission; the Lac du Flambeau band of Lake Superior Chippewa Indians; Steven Fick; the Burns family—Phil, Clint, and Andy—and friend David Wajda; and most especially to the Bennetts—Mike, Vicki, Jaimie, Chuck, and Asa—and the Goldsworthys and everyone at Thunder Lake Marsh—Charlie, Bette, Mark, Denise, and Tim—for their hospitality, and for their patience in answering unending questions. Also thanks to James Kniprath and others at the Bennett Cranberry Company, Phil and Mary Brown, Don Hastings, Donna Fairbanks, the Graybow family, Pamela Walker, and everyone else who helped in so many ways with the text or the photography of this book. Thank you, all!

Carolrhoda Books, Inc. c/o The Lerner Group
241 First Avenue North, Minneapolis, Minnesota 55401

LIBRARY OF CONGRESS CATALOGING-IN-PUBLICATION
Burns, Diane L.
Cranberries: fruit of the bogs / by Diane Burns : photographs by Cheryl Walsh Bellville.
p. cm.
Includes index.
ISBN 0-87614-822-4 (lib. bdg.)
1. Cranberries—Juvenile literature. 2. Cranberries—Wisconsin—Juvenile literature. [1. Cranberries.] I. Bellville, Cheryl Walsh, ill. II. Title.
SB383.B84 1994
633.76′09775—dc20 93-29620
CIP
AC

Manufactured in the United States of America

1 2 3 4 5 6 - I/MP 99 98 97 96 95 94

For my parents, who first showed me Wisconsin's outdoor wonders.
—DLB

For our friends at Marcy Open School and for teachers everywhere. You make a difference.
—CWB

Contents

Ibimi 5
The First Farms 10
The Beds 13
The Planting 19
The Sprinkling and Weeding 23
The Growing and Ripening 28
The Harvesting 33
The Processing 41
Season's End 43
Glossary 47
Index 48

Ibimi

Long before European settlers arrived in North America, a wild red fruit dotted the boggy places of the north Atlantic coast each autumn. This red fruit was called **ibimi** [ih-bih-mee], or "bitter berry," by the native Algonquin people who lived along the shore.

When the European pilgrims arrived, some of these Native Americans shared the ibimi with the newcomers to help them survive the long, cold winter. The pilgrims enjoyed the tart fruit, which was rich in vitamins C, A, and B. Since they had eaten only dried meat and bread during their many months at sea, the ibimi tasted especially good to them.

The arched pink-white ibimi flowers reminded the pilgrims of the neck, head, and beak of a common European bird, the crane. Because of this, the settlers began to call ibimi by another name—"crane-berry." Over time, the word was shortened to cranberry.

The Precious Berry

Cranberries were so important to the pilgrims that they wrote laws protecting the ones growing near their settlements. No picking was allowed in public bogs, and only ripe cranberries could be picked on private land. In this way, they tried to make sure that enough wild cranberries would grow back each year.

New England whaling and trading ships carried barrels of cranberries on board to protect their crewmen against scurvy—an illness caused by a lack of vitamin C. Explorers treasured the bitter berry too. As early as the 1600s, French voyageurs wrote in their journals about finding the fruit. Sometimes they traded for the berries with the Native Americans they met. During the winter of 1805-06, the famous explorers Lewis and Clark purchased dried cranberries from the peoples on the Pacific Coast. Years later, Oregon settlers picked cranberries in the bogs near their homes and sold the fruit to logging and mining camps in California.

Many northern Native Americans still use cranberries in their traditional cooking. They sweeten the fresh berries with sugar, honey, or maple syrup. Sometimes they mash the berries with cornmeal and bake the mixture as bread. Traditionally, ibimi was boiled with venison, wild onions, and fat, and then dried as a food for trips or winter storage. This mixture was often called pemmican.

Throughout history, there have been other uses for ibimi. The bitter maroon juice dyed blankets and utensils red-brown. Mashed, unripe ibimi helped heal scrapes and sores. And it soothed more than just physical injuries. In some Native American nations, the berry was a symbol of peace, helping to heal disagreements.

Donna Fairbanks helps her grandson Ben make a pie.

The cranberry is a creeping woody vine that forms a dense green mat wherever it grows. Cranberries grow naturally in the cool, northern **bogs** of North America, Asia, Europe, and Greenland. A bog is a low area where rain becomes trapped and cannot drain away. Over hundreds of years, piles and piles of dead plants fall to the bottom of the bog, pressing down upon each other. These packed-down plants turn into a spongy, soil-like layer called **peat.** Because peat holds moisture, cranberries grow well in it.

Cranberries like a cool growing season, but they need protection from severe cold. In wild bogs, this protection comes from nearby lakes and ponds that often fill up from spring and fall rains and overflow into the bogs. This extra water covers the cranberry plants, protecting them like a blanket from the freezing cold.

To make a wild bog into a cranberry farm, trees and shrubs and the top layer of peat are removed. Then a layer of sand is dumped over the area, and ditches are dug around it.

The First Farms

Americans first discovered they could grow cranberries as a crop in the 1800s. Early cranberry farms started along the Atlantic coast, from the province of Quebec in Canada south to New Jersey. In 1816, Captain Henry Hall transplanted some wild cranberry vines from one bog along the Massachusetts coast on Cape Cod to another.

This was the first attempt to farm the "bog ruby," as Cape Codders called their cash crop. Soon, other seacoast farmers adopted this idea, and the cranberry industry was born.

As settlers moved westward, they brought their knowledge of cranberry farming with them. From the Atlantic coast to the Midwest to the Pacific Northwest—early cranberry growers shaped their farms with back-breaking labor. They cleared bogs, dug canals, and planted acres of vines by hand.

At first, the farmers grew native American cranberry vines, which bear the scientific name *Vaccinium macrocarpon.* Over the years, however, they mixed different kinds of vines and developed the many types of cranberry plants grown today. These plants have names such as Ben Lear, Crowley, Early Black, Howes, McFarlin, Searles, and Stevens. Some of these cranberry plants grow six feet in length.

Collecting cranberries was difficult for early growers. Workers wrapped strips of cloth around their fingers to protect against cuts and scratches from the plants. Then they crawled across the matted cranberry plants on hands and knees to pick the fruit. Many years later, workers began to use handheld scoops and rakes with wooden teeth to comb the berries from the plants.

In the 1940s, when many workers went overseas to serve in World War II, machines began to replace harvesting by hand. Now, most cranberry farmers use machinery to grow and harvest their berries.

In North America, cranberry farms are now mostly found in four Canadian provinces (Nova Scotia, Quebec, Ontario, and British Columbia) and five of the United States (Massachusetts, New Jersey, Wisconsin, Oregon, and Washington).

Wisconsin and Massachusetts compete each year for the honor of growing the most cranberries in North America. Cranberries are Wisconsin's largest fruit crop. Two of the most well-known cranberry areas in the state are Cranmoor Township in Wood County in the center of Wisconsin and the Thunder Lake Marsh area in the north.

To learn how cranberries are grown and harvested, we'll visit an up-to-date farm in each of these areas, season by season.

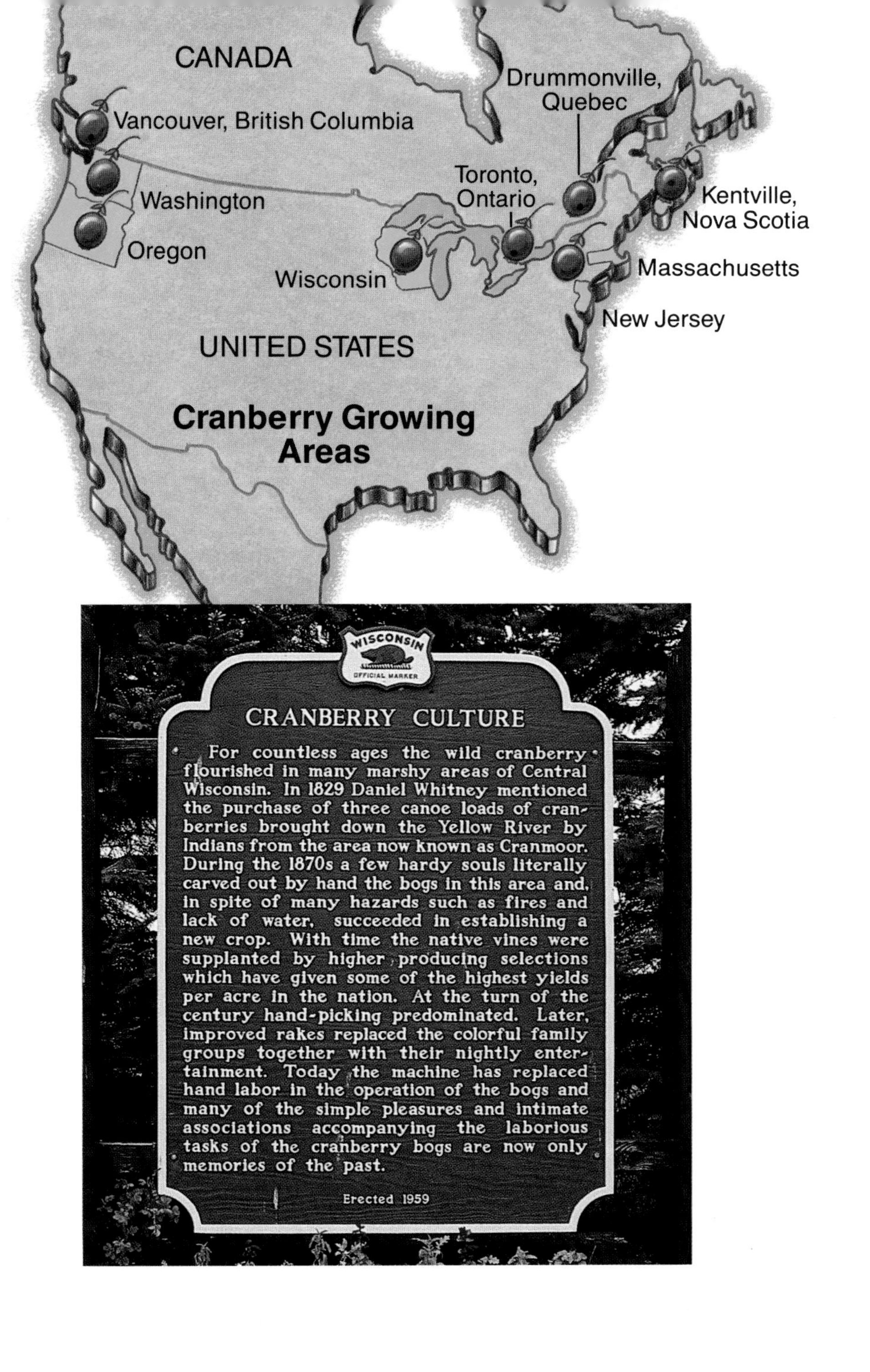

The Beds

In springtime, Cranmoor bogs awaken beneath soft carpets of cranberry-vine green. Along the shoreline of the ditches, herons pose like statues, waiting for small fish and frogs to swim by. Water lilies and blue flags soften the bogs' edges, where ducks and geese paddle about with their young. Welcoming the new season, sandhill cranes rattle the air with their deep voices.

One of the cranberry bogs here belongs to Bennett Cranberry Company, owned by Mike Bennett and his family. They farm 130 acres of cranberries in 60 rectangular areas called **beds.** Each has a sand-peat bottom and is from 2 to 5 acres in size.

The beds lie in an orderly pattern, lined by heaped-up banks of dirt, called **dikes,** and connected by ditches and roads. The ditches have **flow gates,** which control the amount of water in them.

A thousand acres of marshes surround the many beds. The marshes serve as **reservoirs,** or holding areas for fresh water. When the flow gates are opened, the ditches and the beds are flooded with water from the marshes on higher ground. When the flow gates are closed, the water in the beds drains through the ditches into marshes on lower ground.

Mike Bennett learned to grow cranberries as a boy, helping his parents and grandparents. He is part of the fifth generation of Bennetts to work the family bog. Like many Wisconsin cranberry growers, he and his family belong to a cranberry **cooperative.** They sell their annual crop to the co-op, Ocean Spray Cranberries, Inc., which then markets the fruit.

Like the Cranmoor bogs, during March and April, Thunder Lake in the north woods also awakens. The lake is no more than six feet deep. It acts as a 1,800-acre reservoir for the cranberry beds nearby.

The Goldsworthy family has farmed Thunder Lake Marsh since the 1950s. Their cranberries are grown on 125 acres divided into 54 beds. The beds are connected by a system of ditches, underground pipes, and pumping stations. To flood the beds, flow gates in the ditches are opened, and water is pumped into the ditches from Thunder Lake. The water flows from the ditches into the beds. To drain extra water out of a bed, other flow gates are opened, and the water drains into ditches that carry the water back into Thunder Lake.

The Planting

Whether at one of Cranmoor's bogs or at Thunder Lake Marsh, planting new cranberry beds is an important April activity. Growers replant beds that are not bearing enough berries. They also plant new cranberry beds to enlarge their farms.

Each fall, growers clear any areas they plan to plant with new vines. This means pushing old vines off a bed, using a bulldozer. After the cleared ground freezes, truckloads of sand are dumped onto the bed and smoothened.

Now, in spring, the farmers level out the sand more exactly, using lasers. One laser is set inside the scraped bed, another is placed on the blade of a bulldozer. Directed by beeps from lasers in the bed, the driver either adds sand or scoops out sand to make the bed level.

Mike Bennett and workers at the Bennett Cranberry Company mow cuttings (above). Jaimie Bennett displays the cuttings (right).

To plant each bed, 2,000 pounds of cranberry vines are mowed from plants in other beds. These vine pieces, called **cuttings,** are spread out on the new bed. Cuttings aren't fussy. Whichever end of the cranberry vine is planted will take root and grow.

On some cranberry farms, like Mike Bennett's, a tractor pulling a **spreader** is used. Vine cuttings are piled into the box and dropped a few at a time onto the bed. Another machine, the **mechanical planter,** is pulled or pushed behind the spreader. Its round pieces of metal roll and push the cuttings into the ground as they turn.

In four to five years, a new bed will be ready to harvest for the first time, and it will bear cranberries for a hundred years or more.

A spreader (above); a mechanical planter (top right); and newly planted rows of cranberry cuttings (lower right).

Asa Bennett points to the newly planted cuttings, which look like little trees.

The Sprinkling and Weeding

Throughout April, the cranberry **runners,** the woody stems that grow lengthwise across the top of the bog, burst into growth. Millions of **buds** swell up on the runners. The buds are small, red-green coils of shiny leaves, which will grow into upward shoots.

These new shoots are called **uprights,** because they grow up toward the sun. While the buds grow and lengthen into uprights, the cranberry beds become a moist and spongy carpet.

A fully grown runner with uprights and two ripe cranberries

Chuck Bennett shows a friend their radio frost alarm.

All through the growing season, frost is a constant danger. Whenever air temperatures drop below 32°F, the plants can freeze and die. To prevent this damage, cranberry growers use sprinklers. Sprinklers cover the plants with a mist of water. Water gives off heat as it freezes, and this heat keeps the plants warm. As long as the plant is covered with a mist, it won't freeze. The water freezes instead of the tender plant.

Sprinkler systems are assembled early each spring. While connecting miles of pipe sections, the work crews clean out leaves and dirt, and they repair any pipes that are broken. They test the sprinkler pumps to make sure they are working properly. They also place thermometers in the bog's cold spots.

In Cranmoor, Mike and his workers know from experience where to find these cold spots. Their thermometers are wired to radio alarms and a telephone line to Mike's house. Whenever the bog temperature drops near 32°F, the alarms sound and the sprinklers are switched on.

Growers who don't have alarm systems, such as the Goldsworthys, spend hours in their marshes checking their temperature gauges. If they see temperatures dropping into the danger zone, they flip on the switch for the sprinklers.

The worst weeks for springtime frost usually come in April and May. But even after these months have passed, cranberry growers keep a watchful eye on the beds.

Charles Goldsworthy gathers a sample of hardhack, a bog weed (left). The left side of this bog has been weeded and the right side has not (below).

During May and June, cranberry dikes need mowing to keep weeds down. On some cranberry farms, flocks of tame or wild geese help the grower by eating weed seeds. In many bogs, unwanted plants are weeded by hand. Growers may apply weed-killing chemicals on any plants that shade the cranberry vines. Weed control goes on all during the growing season.

In these ways, young cranberry plants are carefully tended as springtime fades into the growing season of summer.

The Growing and Ripening

In early June, starlike cranberry blossoms cover the beds in a pink-white cloud. Growers hire beekeepers to bring hives to their bogs—about one hive for every two acres of blossoming plants. Trucks with forklifts carefully set the beehives in place. As the bees sip the cranberry flowers' sweet juices, they spread pollen from plant to plant, one blossom at a time.

Mowing the dikes is especially important now. Mowing gets rid of the weeds so that bees go straight to the cranberry blossoms.

After a few weeks, the blossoms wither, and the beehives are removed. Waxy green nubbins, called **pinheads,** remain where the blossoms once were—about three per upright. These pinheads will become berries.

By late June, the pinheads are pea-sized. In July, they are the size of marbles. Sprinklers still protect the young crop against any sudden frost. The sprinklers also provide moisture.

During this stage, growers watch their crop for signs of a pest called the fruitworm. The fruitworm burrows inside a young, green cranberry and eats the fruity center—then it goes on to the next berry.

To destroy the worm, cranberry growers must catch it between berries. Knowing all about the fruitworm helps farmers decide exactly when the worm will move from one berry to another. Only then can it be destroyed with chemicals.

This is an example of the damage a fruitworm can do to the cranberry fruit.

During the growing season, many cranberry farmers use chemical fertilizers. These fertilizers, which provide additional food for the growing cranberry plants, are applied by hand or mechanical spreader.

In August, the berries reach their full 3/4-inch size. They are beginning to blush a warm pink.

Then cool September days arrive, turning the leaves of the birches, oaks, and maples into bright yellows, warm golds, and blazing reds. The cranberries darken to a deep, rich crimson. To many people, the sight of these brilliantly colored bog rubies marks the beginning of autumn.

Zack Goldsworthy holds up a ripe cranberry.

The Harvesting

Usually by mid-October, the year's cranberry crop is fully ripe—about 16,000 pounds at every cranberry bog across Wisconsin. It's harvest time!

For some people, this means celebrating, and for others it means long hours of work.

Many communities in cranberry areas, such as the town of Eagle River, near Thunder Lake, celebrate the berry harvest. Eagle River's festivities are held the first weekend in October. Despite the often chilly weather, the weekend's activities warm Cranberry Fest crowds. Festival visitors tour the Goldsworthys' farm at the Thunder Lake Marsh for an up-close peek at a cranberry harvest (below).

Visitors sample cookies, pies, cheesecake, strudel, and fritters, washing them down with cups of hot, spiced cranberry punch.

Work crews dress in hip boots and rain gear for the four-week, twenty-four-hour-a-day job of collecting the cranberries. They remove the sprinkler systems to make room in the beds for harvest machinery. Without the sprinklers, the crop is unprotected against frost.

The crews work in shifts to flood the beds one at a time. In a few hours, several inches of water fills each bed. The berries are now ready to be collected by a method called **wet harvesting.**

To wet harvest ripe cranberries, workers drive **water reels,** called "eggbeaters," through each cranberry bed. The machines move through the bog at about two miles per hour, and their spinning reels churn the water and knock the berries off the vines.

The water reels have wide tires. Able to ride over the beds without sinking, wide-tired machines cause less damage to the cranberry plants than narrow-tired machines. In addition, the mechanical parts of the eggbeaters are greased with vegetable oil. If the oil leaks, it will not harm either the berries or the bed.

Loosened by the churning of the water reels, the ripe berries float to the top of the water. They float because there are small air pockets in each berry.

As the berries are freed from their stems, they spread out into bunches that cover the bog in jigsaw patterns.

To collect these ripe cranberries, workers assemble miles of a yellow, rubberized floating tube, called a **boom.** Other growers use a boom of connected wooden slats or foam-covered chains.

Workers surround the berries in the bed with the boom. Pulling one end of the boom to tighten it, the workers corral the cranberries into a smaller and smaller space. They use long-handled screens or scrapers to guide the berries onto a conveyor belt that slants up to an open-bed trailer. On the moving belt, streams of shooting water and puffs of air remove leaves and sticks, called **trash.** Cranberries on the belt are cleaned at a rate of about 20,000 pounds, or 200 barrels, an hour. A **barrel** is the standard measurement for cranberries. One barrel is equal to 100 pounds of berries.

Tractors pull the trailers to a loading area where the cleaned berries roll up another conveyor belt into a truck. The fully loaded cranberry truck holds 230 barrels of berries. The truck is then driven to a cooperative or factory.

Berries sold fresh must be firm and unbruised, and harvesting with eggbeaters can bruise the fruit. Bruised cranberries are used to make juices, jellies, and other foods. To harvest berries that can be sold fresh, farmers can either wet harvest with a mechanical rake, which combs the berries from the vines, or farmers can use **dry harvesting.** In this method, the beds are not flooded. The cranberries are picked by rubber-tired machines that ''walk'' across the top of the cushiony bog.

Metal teeth along the front edges of the machines comb the berries from the plants and place them gently in a sack. Workers remove full sacks and empty them onto a metal screen. The fruit falls through berry-sized holes in the screen, while trash gets caught on top. Cleaned berries fall into crates, which are then bundled together with nylon belts. Helicopters lift the crates onto trucks that haul them to the cooperative or factory.

Using helicopters prevents damage to next year's cranberry buds, which are already forming on the plants.

The Processing

Berries that are to be sold fresh are treated differently than ones that will be used in juices, jellies, and other foods. First, the berries are air-dried and sorted by size. They move along onto a sloped conveyor belt called a **sorting mill.** By bouncing the berries, this machine shows which berries are of good enough quality to be sold. The fruit starts at the top of the sorting mill and makes its way down the sloping belt. A sound, ripe berry easily bounces over a number of one-inch-high boards, passing the quality test. Each berry has several chances to jump high enough. The berries that pass the test are sorted once more by hand.

Then the fresh fruit is bagged, either by hand or by machine, and shipped to stores. Damaged fruit that fails to bounce is used to make juices and other foods.

Mike Bennett uses wet harvesting to bring in his crop. At Ocean Spray, he unloads his berries into a huge holding pond. A sample is taken to check their quality, and the information is stored in a computer. The amount of berries brought in and their color make a difference in the price a grower will be paid for the crop.

The Cranmoor area's fruit is sold fresh or made into a variety of foods, such as cranberry sauces and juices. They can also be sold as ingredients in jams, mustards, and teas.

In recent years, dried cranberries, somewhat like raisins, have become popular. The dried fruit is made by adding a sugar solution to raw cranberries as they dry. These dried cranberries can be eaten as a snack or used in cooking.

Molly Goldsworthy enjoys a refreshing glass of cranberry juice.

Season's End

Each year, about 400 million pounds of cranberries are harvested in North America—enough for 1-1/2 pounds for each person in the United States. About 50 percent of the annual crop grows on Atlantic coast farms. Wisconsin yields more than 33 percent of the harvest. Washington and Oregon contribute 10 percent, as does Canada.

After the final truckload of cranberries leaves the Wisconsin bogs, crews spend several days gathering, cleaning, oiling, and storing the pieces of harvesting equipment. This year's crop is safely harvested, and next year's tender buds will be safe beneath a protective layer of ice.

The bogs are quieter now. Some creatures have left for warmer climates. Others have burrowed deep into the reservoirs and ditches to wait for springtime. Lonely winds shake the branches of tamaracks and swamp oaks, sending their last leaves to cover the iced bogs, beds, and reservoirs. Snow frames each exposed stem and branch and blankets rocks and clumps of marsh grass. Everywhere, ice etches lacy patterns against the cold, clear sky.

Though the bogs rest, cranberry growers are still busy. They spread a layer of sand, about an inch deep, on the ice over each cranberry bed. When the ice melts next spring, the sand will sift to the bottom. The sand will anchor new plantings, smother weeds, bury insect eggs, and provide extra support for established plants.

The Health of the Bog

Cranberry growers know that their farming methods affect the health of the bog and the reservoirs around it. The water levels of the reservoirs must be raised and lowered several times during cranberry growth. These changes in water levels are good for some species of plants and animals, and harmful for others.

Growers are required to follow strict conservation laws. The Environmental Protection Agency (EPA) and the Wisconsin Department of Natural Resources (DNR) watch the reservoirs to see how the animals and plants are affected and to find out what can be done to make water changes easy on the species living there. These agencies also check how chemicals are used on cranberry farms and how wisely the bogs are cared for.

In northern Wisconsin, Verdun Cecil, a science teacher at Three Lakes High School, encourages his students to study water samples from nearby reservoirs. Testing the samples, the students learn about water quality. By knowing what chemical components are in the water from year to year, Mr. Cecil and his classes can monitor changes, good or bad.

As winter stretches toward springtime, cranberry experts come to Cranmoor and share new information with Mike and his neighbors about farming. At the same time, the Goldsworthys are learning too—reading new cranberry information that could help next year's crop and keep the bog healthy.

Season in and season out, cranberry growers are hard at work. They think ahead to next spring, to the time when their cranberry bogs will bloom again.

Glossary

barrel: standard of measurement for cranberries. One barrel equals 100 pounds of cranberries.

bed: one portion of a cranberry bog, usually rectangular in shape and two to four acres in size

bog: a type of wetland on which cranberries can be farmed. Cranberry bogs have a peat bottom and acidic soil.

boom: a floating fence used to corral wet-harvested cranberries

buds: tiny balls at tip of growing area that swell on cranberry stems and grow into upward shoots

cooperative: a jointly owned business, organized and operated by its members

cutting: a piece of green, growing cranberry vine, six to eight inches in length

dike: a bank of dirt separating a cranberry bed from a ditch or reservoir

dry harvesting: collecting cranberries from unflooded bogs

flow gates: tubes across ditches that regulate the flow of water in and out of the beds

ibimi: the Algonquian name for the cranberry, meaning "bitter berry"

mechanical planter: a bulldozer with a curved metal front that pushes cranberry cuttings into the ground of a cleared bed

peat: a soil-like layer of dead plants that forms at the bottom of a bog

pinheads: the small round remains of the flowers that have hardened and turned green and will eventually grow into ripe cranberries

reservoir: natural or constructed containment for water, such as a lake, pond, or marsh, that can flow or be pumped onto a cranberry bed when needed

runners: woody cranberry stems from one to six feet in length

sorting mill: a factory machine that checks the quality of fresh cranberries

spreader: a box with a roller that attaches to a small tractor. This piece of equipment spreads cuttings apart so they can be planted.

trash: unwanted plant material that is harvested along with the berries

uprights: new cranberry shoots that grow upward from runners to about three inches in height

water reels: wide-tired machines used to churn water in wet harvesting

wet harvesting: collecting cranberries by flooding beds and churning the water to loosen berries from the vines

Index

beds, 14–15; draining, 15, 18; flooding, 15, 18, 35
bogs, farming, 8; wild, 8

chemical use, 27, 30, 31, 45
cleaning berries, 37, 40
cranberry, description, 8; origin of name, 6; scientific name, 10

environmental issues, 45

farms, North American locations, 12
fertilization, 31
frost prevention, 25–26, 30, 35

growth, of berries, 29–30, 32; of buds, 23

harvesting, 33; amounts, 43; celebrations, 34; dry, 40; preparation for, 35; wet, 35–37

history of cranberries, 6; Native Americans, 5, 7; North American farmers, 10–11

pest control, 30, 44
planting, 19, 20–21; preparation for, 19
pollination, 28
post-harvest work, 43–44, 46
pricing berries, 42

selling berries, 17, 40, 42
sorting berries, 41

uses for cranberries, 7, 42

weed control, methods of, 27, 44; during pollination, 28

Photo Credits

Additional photos courtesy of: Library of Congress, p. 5; Montana Historical Society, Helena, p. 6 (right); The Ocean Spray Cranberries, Inc., Corporate Archives, p. 10, 11 (left); John F. Ryan/Plymouth Copters, Ltd., p. 40; Bessie E. Kmiecik, p. 44; The Ocean Spray Cranberries, Inc., p. 46 (top)

Back Cover

Donna Fairbanks and her grandchildren Ashley and Ben cook with cranberries.